Material Choices

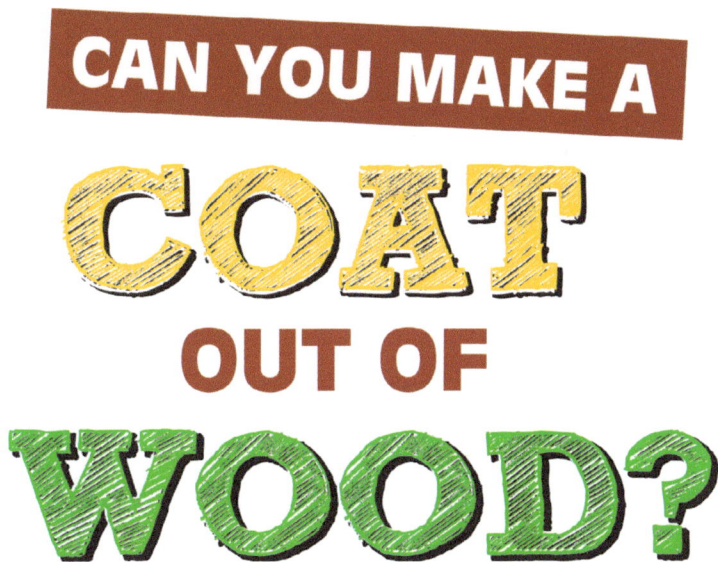

CAN YOU MAKE A COAT OUT OF WOOD?

by Susan B. Katz

PEBBLE
a capstone imprint

Published by Pebble, an imprint of Capstone
1710 Roe Crest Drive, North Mankato, Minnesota 56003
capstonepub.com

Library of Congress Cataloging-in-Publication Data is available on the Library of Congress website
ISBN: 9781666350883 (hardcover)
ISBN: 9781666350944 (paperback)
ISBN: 9781666351002 (ebook PDF)

Summary: Wood has many advantages and is used for everything from building houses to keeping houses warm. Learn more about different types of wood and their many uses.

Editorial Credits
Editor: Christianne Jones; Designer: Elyse White; Media Researcher: Morgan Walters; Production Specialist: Polly Fisher

Image Credits
Getty Images: FatCamera, 19, shironosov, 5; Shutterstock: Africa Studio, (pine) 11, andrey_l, (paper) 21, Andrii A, top 9, B Brown, 12, cmnaumann, (right) Cover, Dan Thornberg, (maple) 11, Elena Elisseeva, (cedar) 11, HP Productions, (cherry) 11, Jesus Cervantes, (eucalyptus) 11, kzww, bottom 9, lovelyday12, 7, mavo, 13, Monkey Business Images, 14, Nik Merkulov, (sticks) 21, NinaMalyna, 10, OlgaGi, (left) Cover, Oxie99, Simlinger, 6, vvoe, (birch) 11, wavebreakmedia, 15, yda Productions, 17

Table of Contents

A Wood Coat? 4

Properties of Wood.......................... 6

Using Wood...................................... 12

Wood History..................................... 16

Wood and Coats............................. 18

Activity... 20

Glossary 22

Read More 23

Internet Sites 23

Index... 24

About the Author 24

Words in **bold** are in the glossary.

A Wood Coat?

Coats are meant to keep you warm. They keep you dry from snow and rain. They protect you from the wind. They have to fit well but not be too tight.

Do you think you could make a coat out of wood? Let's find out!

Properties of Wood

Wood comes from trees and is a **renewable** resource. Trees are chopped down and cut into pieces.

Since trees release oxygen, it's important to replace them. We should plant new ones if we cut them down.

Wood is a hard material. It does not bend easily.

Wood is also rough. It has to be **sanded** to make it smooth.

Some types of wood are very thick and heavy, like oak. Some are thin and light, like birch.

oak

birch

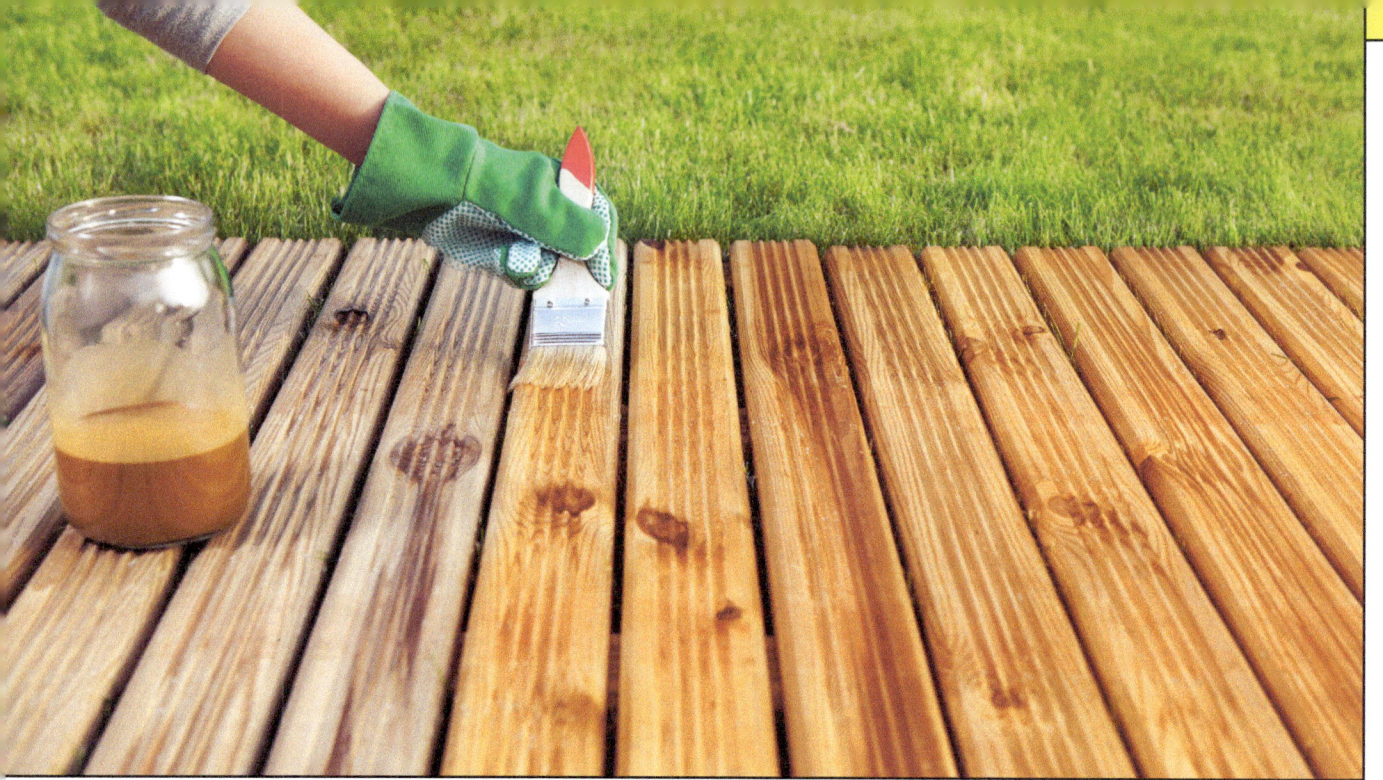

Treating wood keeps it from rotting.

Wood is made up of tiny fibers. Wood often soaks up moisture when it is wet. It has to be **treated** to make it waterproof.

Wood comes in different colors. Cherry wood is dark. Birch wood is light. The color of wood often darkens when it's treated.

Wood has different smells too. Cedar, white pine, **eucalyptus**, and maple all have their very own smells.

| cherry | birch | cedar |
| white pine | eucalyptus | maple |

Using Wood

Wood can be carved into interesting shapes. It can be painted, nailed together, and even bent into a curve. Many houses and other buildings are made out of wood. Furniture is made out of wood because it's hard and sturdy.

People make cabinets, kitchen tables, and dressers out of wood. Bed frames are also made from wood. Many pictures hanging on your wall have wood frames too.

Wood is used to make paper. Every book
you read started as a tree!

Wood History

Wood has always been an important **raw** material. It has been used throughout history. People used it to build fires and stay warm. Today, wood is still used for fires. People also use wood to build shelters. Many buildings are still made from wood.

Wood and Coats

When you wear a coat, you need to move. A wooden coat would be very **stiff**. It would be very heavy too.

A coat needs to keep you warm and dry. Unless wood is treated, it soaks up water. That wouldn't keep you very warm!

So could you make a coat out of wood? Would you?

ACTIVITY

Wooden Name

Wood is part of your everyday life. It is all around you! Go outside and find a bunch of sticks. Then ask an adult to help you make your name out of the sticks. You can hang it on your bedroom door.

What You Need:

- a thick piece of paper

- pencil

- glue gun and hot glue

- sticks

- scissors or wire cutter

- tape

What You Do:

1. Write your name in capital letters on the paper.

2. Have an adult help you cut the sticks to make the letters.

3. Glue the sticks over your name.

4. Let dry.

5. Use some tape and hang your wooden name on your bedroom door.

Glossary

eucalyptus (you-kuh-LIP-tuhs)—a type of fast-growing, evergreen tree that is used for wood, oil, gum, and as a home for nesting owls, crows, and other birds

raw (RAW)—natural

renewable (ri-NOO-uh-buhl)—a resource or material that can be replenished or replaced

sand (SAND)—to polish or smooth with sandpaper or a tool called a sander

stiff (STIF)—hard

treat (TREET)—to add something to protect a material

Read More

Howes, Katey. *Be a Maker*. Minneapolis: Carolrhoda Books, 2019.

Rivera, Andrea. *Wood*. Minneapolis: Abdo Zoom, 2018.

Rustad, Martha E. H. *What Is It Made Of?* Minneapolis: Millbroke Press, 2016.

Internet Sites

50+ Nature Activities
mothernatured.com/nature-play/nature-play-activities-backyard

Britannica Kids: Wood
kids.britannica.com/students/article/wood/277786

Kiddle: Wood Facts for Kids
kids.kiddle.co/Wood

Index

birch, 9, 11

cedar, 11

cherry, 11

eucalyptus, 11

fibers, 10

furniture, 12, 14

history, 16

houses, 12

maple, 11

oak, 9

paper, 15, 20, 21

renewable, 6

treating, 10, 11, 18

trees, 6, 7, 15

waterproof, 10

white pine, 11

About the Author

Susan B. Katz is an award-winning Spanish bilingual author, National Board Certified Teacher, educational consultant, and social media strategist. When she's not writing, Susan enjoys salsa dancing and spending time at the beach. www.SusanKatzBooks.com